Clara Barton

deeply about pe

was the daughter of a farmer. Clara
became a teacher and later a nurse. She desired to help anyone in trouble. She had a gentle touch and a beautiful heart.

The Civil War broke out in 1861. Clara helped each soldier in any way she could. She brought fresh clothes and other supplies to the battlefield. She nursed soldiers' wounds. She did not fear being caught in the line of fire.

Clara slung pots and pans over her shoulder. She would cook meals for a column of soldiers that needed food. Clara served as well as any colonel or lieutenant in the army. People called her "the angel of the battlefield."

After the war, Clara formed a group that looked for missing soldiers. She felt the pain of each wife, mother, daughter, aunt, and cousin who had lost someone. She felt great sadness for losses on both the northern and southern sides.

In 1869, Clara traveled overseas. There she learned about a group called the Red Cross. It was begun six years before to help people in trouble during wartime. Clara saw the beautiful work that the Red Cross did overseas. It managed to touch her heart deeply.

Clara returned home in 1873. She spoke to government leaders about the need for a Red Cross branch in the United States. It took several years before her plan caught on with many people. But slowly, her number of supporters began to climb.

In 1881, a U.S. branch of the Red Cross was started. Clara was its leader. The group helped people in trouble in peace time as well as war. They gave clothes to victims of fires and earthquakes. They nursed coal miners whose cough worsened to pneumonia.

Clara Barton headed the Red Cross until the age of 82. She did everything with love, whether nursing a soldier or fitting a shoe on a needy child. She really was someone with a gentle touch and a beautiful heart.